Crawly

TERMITES

LYN SIROTA

Bolt is published by Black Rabbit Books
P.O. Box 3263, Mankato, Minnesota, 56002.
www.blackrabbitbooks.com

Marysa Storm, editor; Grant Gould, designer;
Omay Ayres, photo researcher

Names: Sirota, Lyn A., 1963- author.
Title: Termites / by Lyn Sirota.
Description: Mankato, Minnesota : Black Rabbit Books, [2020] | Series: Bolt. Crawly creatures | Audience: Age 9-12. | Audience: Grade 4 to 6. | Includes bibliographical references and index.
Identifiers: LCCN 2018019562 (print) | LCCN 2018021642 (ebook) | ISBN 9781680728200 (e-book) | ISBN 9781680728149 (library binding) | ISBN 9781644660256 (paperback)
Subjects: LCSH: Termites–Juvenile literature.
Classification: LCC QL529 (ebook) | LCC QL529 .S57 2020 (print) | DDC 595.7/36-dc23
LC record available at https://lccn.loc.gov/2018019562

Printed in the United States. 1/19

Image Credits

Alamy: Avalon/Photoshot License, 22–23 (top); morgan hill, 26; iStock: Atelopus, 21 (top); PK6289, 8–9; GlobalP, 24 (mammal); Teresa Otto, 14 (btm); Newscom: Mitsuhiko Imamori/Minden Pictures, 4–5; Piotr Naskrecki/Minden Pictures, 21 (btm); Shutterstock: AmyLv, 24 (wood); Andrew Paul Deer, 28–29; Anteromite, 24 (ants); Apisit Wilaijit, 17; Anut21ng Photo, 18; Chaikom, 22; chakkrachai nicharat, 6, 24 (termite); Chipmunk131, 29; Cora Unk Photo, 14 (top); iamporpla, 28; khlungcenter, 32; KiattisakCh, 24 (poop); leungchopan, 27; moj0j0, 12–13 (silhouette), 16; NNphotos, 11 (top); Oleksandr Lytvynenko, 24 (grass); opportunity_2015, 15; Pedristico, 1, 31; PK6289, Cover, 22–23 (btm); plew koonyosying, 3; Protasov AN, 23; Snowboard School, 12–13 (bkgd); Susan Schmitz, 24 (lizard); Sylvie Lebchek, 11 (btm); Syrus Neilson, 11 (middle)
Every effort has been made to contact copyright holders for material reproduced in this book. Any omissions will be rectified in subsequent printings if notice is given to the publisher.

CONTENTS

CHAPTER 1

Meet the TERMITE

Skitter here. Scatter there. A termite nest is full of hustle and bustle. Busy worker termites take care of the queen. Some workers feed it. Others clean it. Meanwhile, other workers repair the nest. They use poop and spit to fix it up. Soldier termites are ready to warn of danger. Their job is to protect the **colony** from attackers.

TERMITE QUEEN
WORKER TERMITES

COMPARING SIZES
worker
about .4 to .6 inch
(1 to 1.5 centimeters)
soldier
about .6 to .8 inch
(1.5 to 2 cm)
king
about .4 to .8 inch
(1 to 2 cm)
queen
inches
1
2

Social Insects

Termite nests are busy places. They're full of pale, blind, and wingless termites. These are soldiers or workers. Only termites that can **reproduce** have eyes and grow wings. They are future kings and queens.

Termites **communicate** using **chemical** signals. Each colony has its own scent. Termites also bang their heads against the walls to warn of danger. Termites inside sense the **vibrations**.

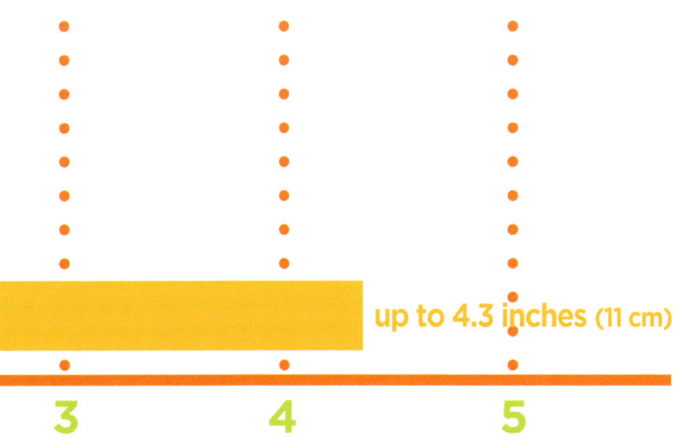

TERMITE FEATURES

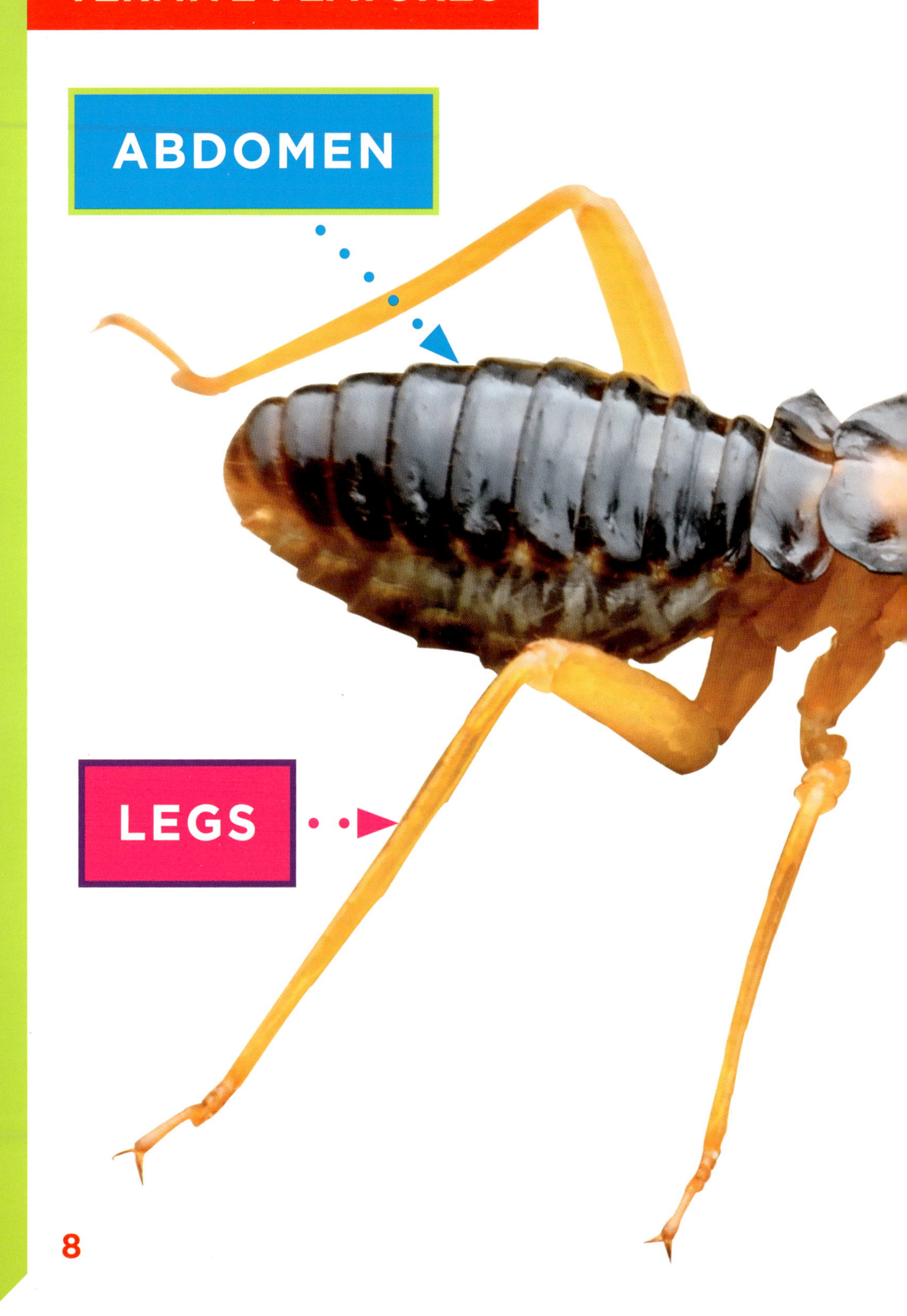

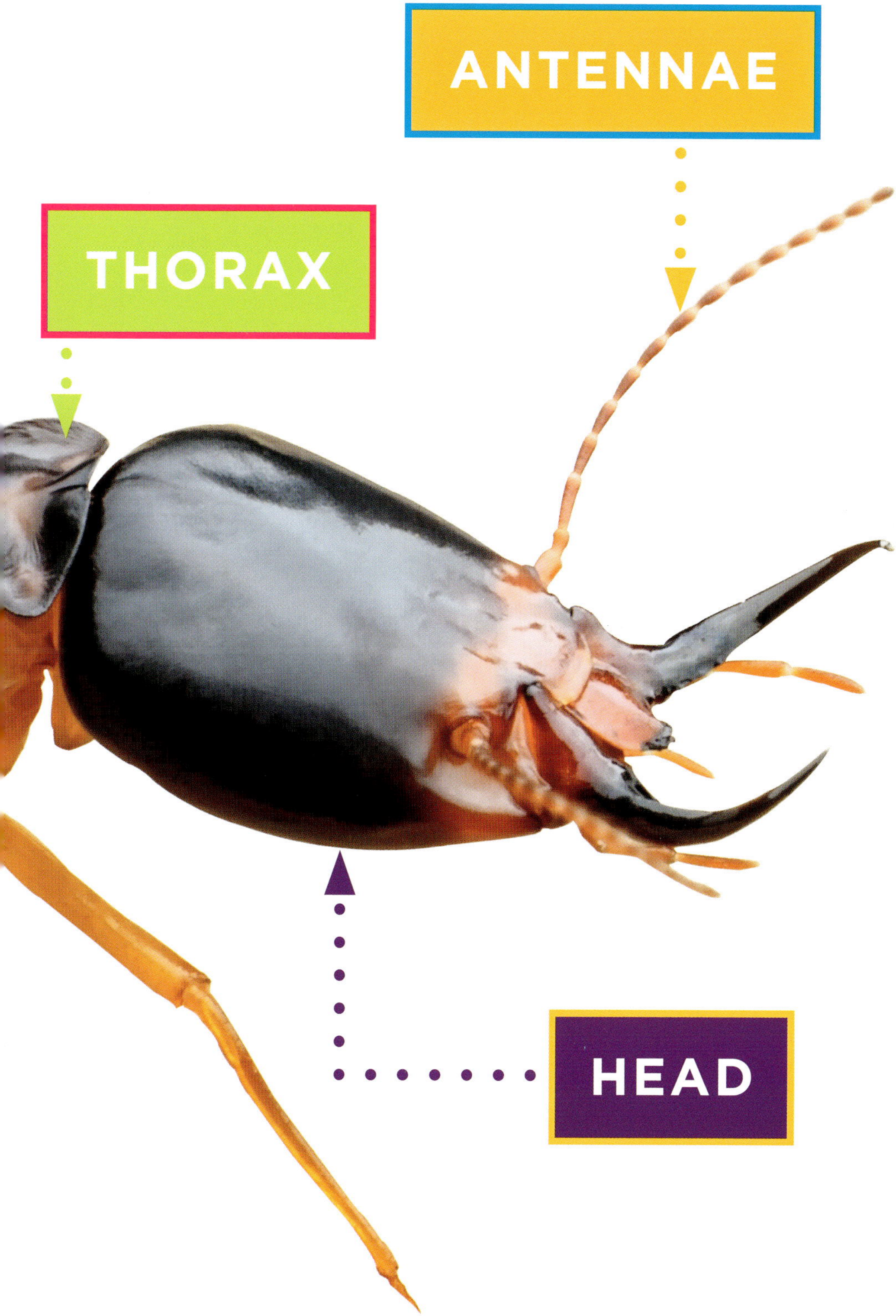

ANTENNAE

THORAX

HEAD

CHAPTER 2

WHERE THEY LIVE and What They Eat

There are many different kinds of termites. They live all around the world. Some live near coasts. Others like forests, deserts, or grasslands. Most termites love moist, **humid** habitats.

Many termites live underground to avoid heat.

FORESTS
DESERTS
GRASSLANDS

TERMITE RANGE MAP

Termites live on every continent except Antarctica.

Different Types of Nests

mounds and hills

nests in wood

underground nests

Nests and Mounds

Termites build fancy homes. Some make nests in the wood they eat. Other nests are poop palaces! Termites make them from soil, spit, and poop. Mounds can rise above ground or be hidden below.

What's for Lunch?

Termites eat dead plants and trees. Some termites like wood that has **fungi** growing on it. Fungi breaks down the wood. It makes the wood easier for termites to **digest**.

A type of bacteria helps termites digest wood. They aren't born with the bacteria, though. Workers must feed young termites their poop to give them the bacteria.

CHAPTER 3

Termites live secret lives. Most never leave their nests. Termites that grow wings take flight in groups. They leave in swarms to make their own colonies. These termites **mate**, find a home, and lose their wings.

Termite Babies

Queen and king termites spend their lives mating. They produce the termites that fill their colony. Eventually, the queen becomes too big to move. Workers and the king take care of it and the babies.

Workers take termite eggs to nurseries. The workers then feed the young once they hatch. Young termites become soldiers, workers, or future queens and kings.

Worker termites only live about two years. Some queens can live for more than 10 years.

Termite LIFE CYCLE

Queens give birth to eggs.

Larvae molt and grow into adults.

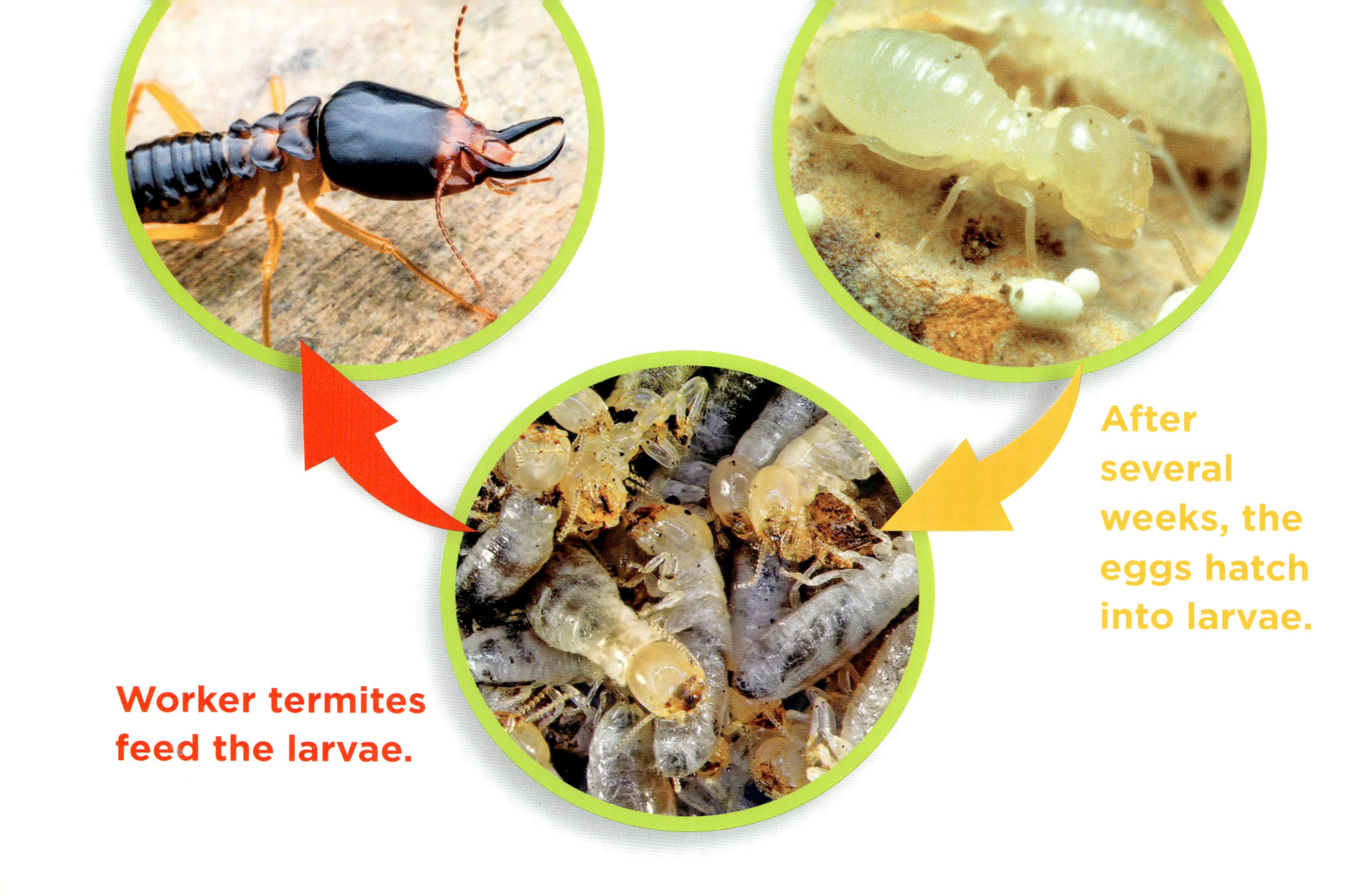
After several weeks, the eggs hatch into larvae.
Worker termites feed the larvae.

Termite Food Chain

This food chain shows what eats termites. It also shows what termites eat.

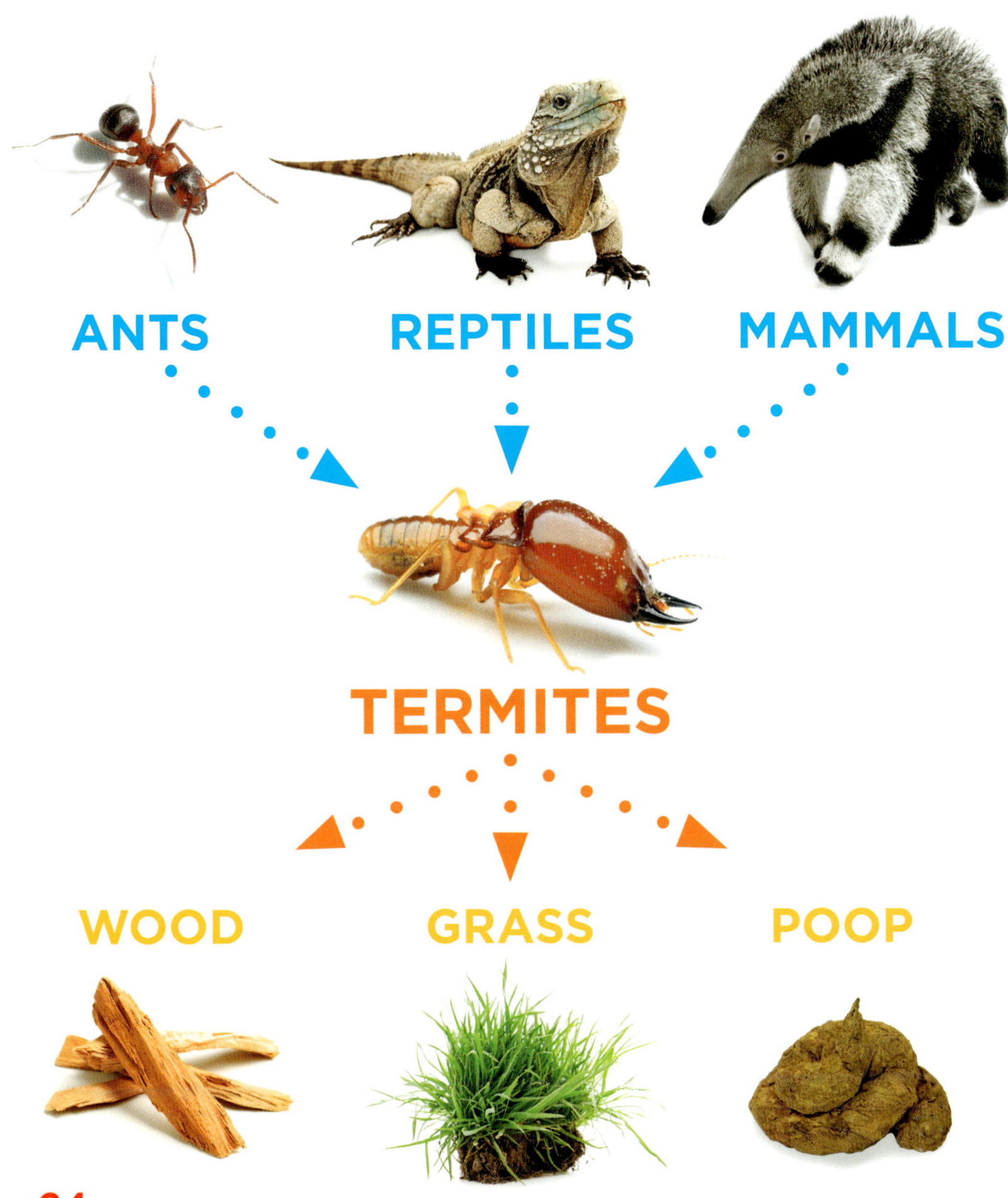

CHAPTER 4

THEIR ROLES in the World

Termites play big parts in their world. One of those parts is as food. Many animals eat termites. Ants march into nests and attack.

Helpful Termites

Termites aren't just food. They're recyclers too. Their bodies break down plant **fiber**. The termites' poop then helps create new and improved soil. Healthy soil means healthy plants and forests.

Some soldier termites spray a sticky substance at attackers. This substance traps attackers.

BY THE NUMBERS

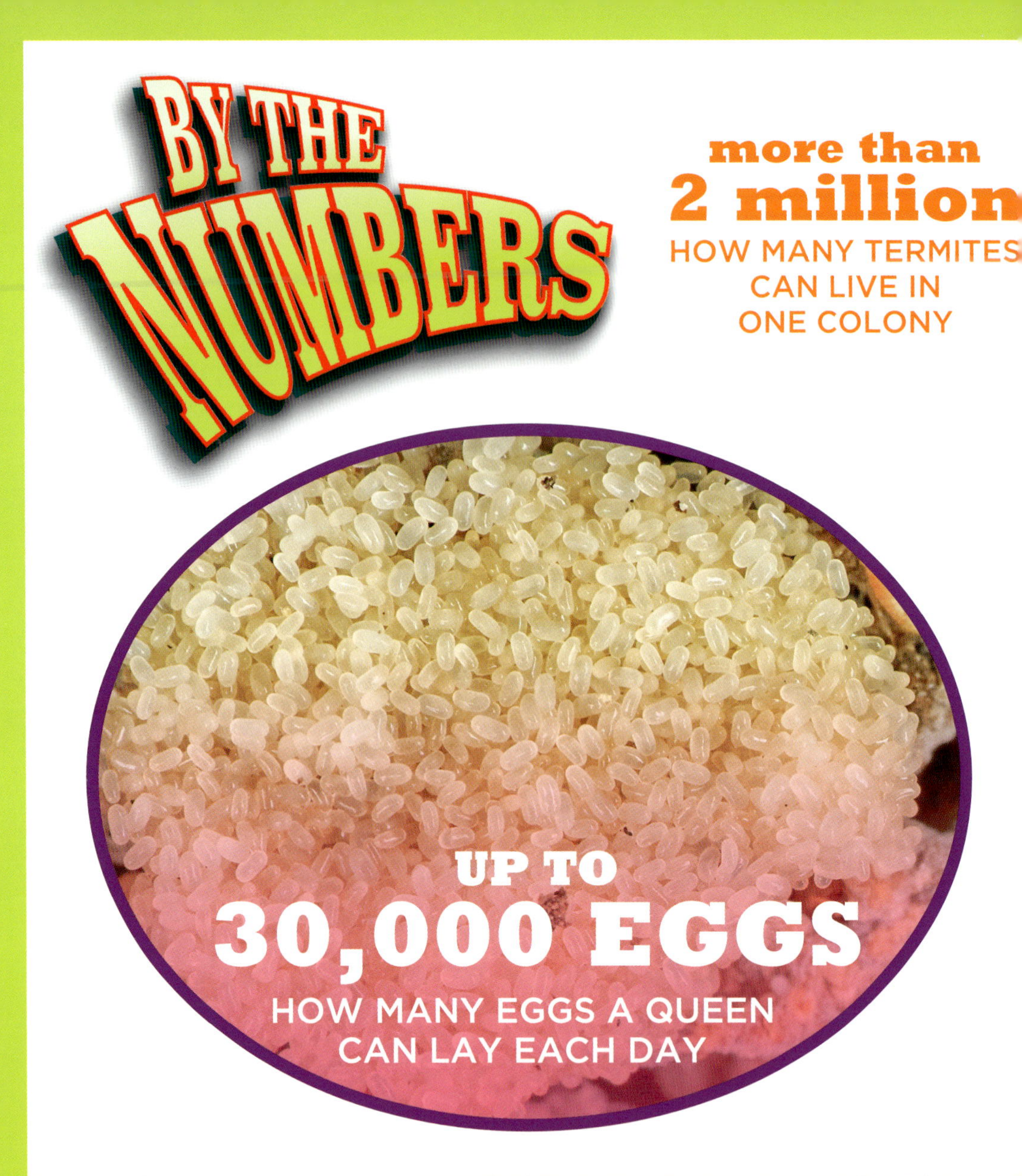

more than 2 million
HOW MANY TERMITES CAN LIVE IN ONE COLONY

UP TO 30,000 EGGS
HOW MANY EGGS A QUEEN CAN LAY EACH DAY

MORE THAN 2,000
NUMBER OF TERMITE SPECIES

HOW HIGH TERMITE MOUNDS CAN BE

bacteria (bak-TEER-ee-uh)—a small living thing

chemical (KEH-muh-kuhl)—a substance that can cause change in another substance

colony (KAH-luh-nee)—a group of animals of the same type living closely together

communicate (kuh-MY-nuh-kayt)—to share information, thoughts, or feelings so they are understood

digest (DY-jest)—to change the food eaten into a form that can be used by the body

fiber (FI-burh)—mostly indigestible material in food

fungus (FUN-gus)—a living thing, similar to a plant that has no flowers, that lives on dead or decaying things

humid (hyoo-MID)—having a lot of moisture in the air

mate (MAYT)—to join together to produce young

reproduce (ree-pruh-DOOS)—to produce new individuals of the same kind

vibration (vahy-BREY-shuhn)—a quick motion back and forth

BOOKS

Lawrence, Ellen. *Building with Poop*. The Scoop on Poop. New York: Bearport Publishing, 2018.

Patterson, Jack K. *Termites.* Creepy Crawlers. New York: Cavendish Square, 2019.

Stewart, Amy. *Wicked Bugs: The Meanest, Deadliest, Grossest Bugs on Earth.* Chapel Hill, NC: Algonquin Young Readers, 2017.

WEBSITES

Termite
kids.nationalgeographic.com/animals/termite/#termite-queen.jpg

Termites
pestworldforkids.org/pest-guide/termites/

Termites – Insects that Eat Wood
easyscienceforkids.com/all-about-termites/

INDEX